Making Microtubules Glow: A Cell and Molecular Biology Laboratory Manual; First Edition

Copyright © 2015 by Thomas Mennella

Additional resources are available on the companion website: http://thomasmennella.wix.com/mtglow

Note to all readers:
Safety during all laboratory exercises is of paramount importance. The safety of the students using this laboratory manual is the sole responsibility of the instructor of the laboratory course and the institution offering that course. The author of this manual bears no responsibility for such safety and cannot be held responsible for any injury, or similar incidents, that may occur while using this manual.

TABLE OF CONTENTS

Lab Exercise #1:

Yeast Genomic DNA Preparation

Lab Exercise #1: Preparation of Yeast Genomic DNA

Background: In order to make many many copies of the *TUB1* gene (which is the gene carrying the instructions for building the tubulin protein; one of the building blocks of microtubules), we need 'template DNA'. Just as you cannot copy a single page of a textbook without the original textbook, you cannot copy the *TUB1* gene without an original *TUB1* gene. For PCR, which is our DNA Xerox machine, the template DNA contains the targeted sequence for amplification; in our case the *TUB1* gene. We must prepare template DNA that contains the *TUB1* gene from yeast in order to PCR amplify it. To do this, we must prepare total genomic DNA from yeast cells.

Objective: To prepare yeast genomic DNA suitable for PCR.

Protocol
The following steps were done for you by your instructor in the interest of time:
- wildtype yeast cells were grown overnight in a shaking water bath at 30ºC in rich media (YPD)
- cells were centrifuged at low speed for 5 minutes and the media (supernatant) removed
- cells were washed with 25 mL of zymolyase buffer (washing consists of resuspending the cell pellet in the buffer and then repeating the above spin and supernatant removal)
- cells were resuspended in zymolyase buffer and aliquoted into 500uL samples

1) to the cells add 100uL of zymolyase (10mg/mL)
    - and 1uL of β-mercaptoethanol (BME ***STINKS*** – add BME under a hood)

2) mix well by inverting

3) incubate at 37ºC for 30 minutes
    - *in this time the zymolyase will destroy (chew up) the oligosaccharide cell wall of each yeast cell. This cell wall is very durable and protects yeast from the harsh environment. To burst yeast cells ('lyse') and liberate their DNA, this cell wall must first be removed. BME breaks disulfide bonds in the proteins of yeast's cell membrane also helping to lyse cells.*

4) spin your samples down – ~10,600 RCF for 4 minutes

5) remove the supernatant with a pipetteman (P1000 set to ~700uL)

6) resuspend cell pellet in 500uL of lysis buffer
    - *the pellet is sticky. No cell wall means the fats of the cell membrane are sticking together. Resuspend the pellet by pipetting up and down; your instructor will demonstrate.*

7) incubate at 65°C for 30 minutes

   - the lysis buffer contains detergent which will dissolve the fats of the yeast cell membrane (much like dish detergent dissolves the fats on your dirty plate). This detergent, combined with high heat, bursts the yeast cells. The lysis buffer also contains a pH buffer (TRIS in this case) and a chemical that protects the genomic DNA (EDTA – to be discussed now...)
   *- **any questions...? now's a good time!***

8) add 200uL of 5M Potassium Acetate – mix well by inverting

9) transfer your tubes to ice and incubate on ice for 20 minutes

   - the detergent in the lysis buffer is necessary to burst the yeast cells, but it's bad news for PCR and/or any other applications we'd like to use our genomic DNA for. So, we must get that detergent out of there. The positive potassium ions, from the dissolved potassium acetate, bind to the negative detergent ions and form a salt. This salt is insoluble and falls out of solution as a white precipitate. You'll know if this step worked correctly if you see a white precipitate form.

10) spin your samples down ~20,000 RCF for 5 minutes

11) transfer the supernatant to a clean tube (throw the white pellet away)
   - **the DNA we want is still in solution – in the liquid – so KEEP THIS LIQUID!**

12) Remove 200uL of your supernatant sample from Step 11 and transfer it to a new tube

13) add 600uL of GX1 buffer to the 200uL of sample from Step 12

14) mix the GelElute ('GE') well and add 15uL of GelElute ('GE') to your tube from Step 13

15) incubate the sample at RT, mixing frequently by inverting, for 5 minutes

16) centrifuge the sample for 30 s (low speed) and carefully remove and dispose of the supernatant

17) wash the pellet with 500 µl Buffer GX1 (i.e., resuspend the pellet by vortexing. Centrifuge the sample for 30 s, then remove and dispose of the supernatant)

18) wash the pellet **_twice_** with 500 µl Buffer GE. (i.e., resuspend the pellet by vortexing. Centrifuge the sample for 30 s, then remove and dispose of the supernatant)

19) air-dry the pellet for 10–15 min or until the pellet becomes chalky white

20) to elute DNA, add 75 µl of TE and resuspend the pellet by pipetting up and down

21) incubate at 50°C for 5 minutes

22) centrifuge samples for 30 s

23) carefully pipet the supernatant into a separate, clean and **_labeled_** tube. The supernatant contains the purified DNA.

These samples will be stored at -20°C (a standard freezer) until your next lab meeting

Reflection Questions:
1) What would be the expected charge on the Gel Elute resin used to purify the DNA in Step #14?

2) TE is used to elute the DNA from the gel elute. TE is a basic solution. How might pH be involved in DNA elution?

Lab Exercise #2:

Polymerase Chain Reaction (PCR) and Agarose Gel Electrophoresis of TUB1

Lab Exercise #2: Polymerase Chain Reaction (PCR)

Background: We need to generate a large number of copies of the *TUB1* gene so that we can 'clone' it into a bacterial pseudo-chromosome (called a "plasmid") containing the DNA sequences for the Green Fluorescent Protein (GFP). We will use PCR to make these copies of *TUB1*. We have our yeast genomic DNA (to serve as template in our PCR) and so today we must set up our polymerase chain reactions.

Objective: To PCR amplify the *TUB1* gene from yeast genomic DNA.

Protocol

1) make a "master mix" for PCR
If your PCR 'works' and you see DNA, how do you know you amplified TUB1 from yeast...? Maybe another microscopic fungal spore fell into your tube while you were setting up the reaction... To be sure we're amplifying what we want, we will do a "NO DNA CONTROL". This reaction will be identical to our real sample, except it will lack the template DNA. If we amplify TUB1 in the NO DNA CONTROL, then we've got problems. If you put nothing on the scanner bed of the Xerox machine and you still get a copy of a page from a book, something's wrong. So, we'll set up two identical reactions by making one big reaction (twice as much as we need) and splitting it in half just before adding the template DNA.

> **In a PCR tube (thin-walled) add the following:**
> 75 uL water
> 20 uL of 5X PCR buffer
> 2uL of dNTPs
> 1uL of forward primer (#1)
> 1uL of reverse primer (#2)
> 1 uL of Taq enzyme

2) mix well by pipetting up and down
the Taq polymerase is stored in glycerol to protect it. This makes it heavier than water. Without mixing the reaction, the Taq will be at the bottom of the tube and the reaction will not work.

3) remove 48 uL of your PCR master mix and transfer into another PCR tube

4) label your two PCR tubes '#1' and '#2' and include your initials

5) add 2uL of your genomic DNA from last week to tube #1
   - add 2uL of water to tube #2 (this is your NO DNA CONTROL)

6) place your samples in the PCR machine (thermocycler)
    - your instructor will set up the thermocycler with a program such as:

    98°C for 30 seconds – repeat 1X
    ⎡ 98°C for 5 seconds
    ⎪ 60°C for 5 seconds   ⎫ repeat 36X
    ⎣ 72°C for 1 minute and 15 seconds   ⎭
    72°C for 5 minutes – repeat 1X

The DNA will now be copied/amplified: at 98°C the dsDNA will separate, then the primers anneal at the 60°C step, and at 72°C Taq polymerase elongates a new strand of DNA starting at the primers. Do this 36 times and you've got a lot of TUB1 genes to work with.

Your instructor will take these samples out of the thermocycler once the reaction is complete.

7) take the pre-run PCR samples that your instructor completed earlier

- transfer 20uL of your reaction to a new, clean microfuge tube and add 4uL of 5X agarose gel loading buffer to it (each lab group should have one sample)

- return the remaining 30uL of PCR product (still in its original PCR tube) to your instructor

- • *your PCR samples are essentially water, as is the 1X TAE buffer that the gel is submerged in. When you add water to water it disperses. We want our sample to collect at the bottom of the well in the gel so that our DNA runs into the gel once a voltage is applied. To accomplish this we make our sample heavier than water (literally). The agarose gel loading buffer is simply 5X TAE with a lot of glycerol and some blue dye. The glycerol makes the PCR samples heavier than water so that these DNA samples immediately sink in the 1X TAE that the gel is in, the blue dyes let us track our samples as we load the gel and as they run in the gel.*

8) load 20uL of your sample in the gel. (your instructor will demonstrate)
    * MAKE NOTE OF WHICH WELLS YOU LOADED!!! *

9) the gel will now run for ~1.5 hrs at 80 volts
DNA is negative. The gel is oriented so that the negative charge is above the wells and the positive charge is below. The DNA is repelled by the negative charge and attracted to the positive. So, the DNA runs down the gel. The gel is a matrix of pores. The DNA must wiggle through these pores to run down the gel and get to that positive charge. Smaller DNA fragments wiggle better than larger ones and so they run faster. Therefore, agarose gels separate DNA based on their size or length ("SIZE FRACTIONATE"). This is also an indirect method for purification…

* WEAR GLOVES – the next steps may involve a CARCINOGEN *

10) Place the gel on the UV light box. Close the lid and turn the light box on.
Find where your samples are loaded. Do you see a band? _____

This agarose gel has been 'cast' in the presence of ethidium bromide (EtBr). EtBr does two things: it "intercalates" between the bases of DNA. This is why and how it is a CARCINOGEN! As DNA runs through the gel it picks up more and more EtBr that comes along for the ride. But, I said TWO things! EtBr also fluoresces under UV light. So, you've got DNA carrying tons of EtBr under UV light... the DNA glows. It glows orange, in fact. This is how agarose gels, along with EtBr, let you see – with you naked eye – DNA!

11) Print out a picture of your gel to include as part of your results documentation.

Reflection Questions:
1) Why might ethidium bromide be a carcinogen?

2) Taq polymerase requires magnesium ions as a co-factor, as do many DNA-binding proteins. Why might this be?

Lab Exercise #3:

Purifying the TUB1 PCR Product and Restriction Enzyme Digestion

Lab Exercise #3: Restriction Digests to Prepare for Cloning

Background: one way or another, we amplified and isolated *TUB1* genes. We can now prepare our DNA for cloning. To prepare DNA for cloning, we cut it. Using restriction enzymes, we essentially make the DNA into jigsaw puzzle pieces that will only fit together the way we want them to.

Objective: To prepare our *TUB1* and plasmid DNA for cloning using restriction endonuclease digestions.

Protocol
PCR DNA Purification using Minicolumn

1) Add 80µl of water to remaining ~30uL of PCR product from the previous exercise

2) Add 500µl GX1 Buffer to your sample and mix well.

3) Be sure your column is nested into a flow-through tube. Transfer the entire sample into the spin column, and centrifuge at 1700 RCF for 1 minute. Discard the flow through liquid.

4) Add 500µl GE Buffer to spin column, and centrifuge at ~15,300 RCF for 1 minute. Discard the flow through liquid.

5) Add 500µl GE Buffer to spin column, and centrifuge at ~15,300 RCF for 1 minute. Discard the flow through liquid. (Yes, this is a repeat of the previous step to achieve two washes of the column)

6) Centrifuge again for 2 minutes (with the column empty) to remove the residual ethanol.

7) Discard the flow-through tube. Set the column into a new, clean, labeled microcentrifuge tube.

8) To elute DNA, add 25uL of TE onto the membrane and let stand for 2-5 minutes.

9) Centrifuge at ~15,300 RCF for 1 min to elute the DNA. Discard the column.

Restriction Enzyme Digest

1) add the following to a new clean microfuge tube:

    10uL water

    10uL of your cleaned PCR DNA of *TUB1*

    2.5 uL of 10X enzyme buffer

    1 uL of Kpnl   these are your restriction enzymes;

    1 uL of BamHI   they will do the cutting

2) set up another reaction in a clean microfuge tube by adding the following:

    19 uL of water

    1 uL of plasmid DNA

    2.5 uL of 10X enzyme buffer

    1 uL of Kpn1

    1 uL of BamHI

The restriction enzymes are going to cut the bacterial pseudo-chromosome (plasmid) and the TUB1 *gene in the same way leaving "sticky ends". These are the jigsaw pieces I mentioned above. The* TUB1 *gene will then fit into the plasmid perfectly. This is cloning.*

3) vortex both samples briefly

4) spin down for ~10 seconds at ~10,500 RCF
simply to collect the entire reaction at the bottom of the tube

5) incubate at 37°C for ~2 hours

- your instructor will collect and freeze your samples after this incubation is complete.

Reflection Questions:

1) Explain how DNA that has been cut with restriction enzymes can fit with other cut DNA fragments just as puzzle pieces fit together.

2) What would be the expected consequences of incubating the restriction enzyme digestion reaction at 95°C instead of 37°C?

Lab Exercise #4:

Gel Purification of TUB1 and GFP Vector - Ligation

Lab Exercise #4: Purification and Ligating of Restriction Digests to Clone *TUB1*

Background: We're finally going to do it! We're going to clone the *TUB1* gene into a plasmid that already contains the GFP gene. After today, we should have our recombinant gene built. The rest of the semester will be spent verifying it and then using it.

Objective: To purify the GFP plasmid and *TUB1* PCR product away from the restriction enzymes and ligate them together as a new recombinant plasmid

Protocol

1) take your RE digest samples from last week and add 5uL of 5X agarose gel loading buffer to them (each lab group should have two samples: PCR product and plasmid)

2) load 25uL of each sample in the prepared gel.
    * MAKE NOTE OF WHICH WELLS YOU LOADED!!! *

3) the gel will now run for ~1 hr at 80 volts

Pour a Gel
*We will practice pouring our own gel to gain this experience. Work in teams of **three**.*

4) set up/tape up a gel tray as demonstrated by your instructor

5) in a 250mL flask, combine:
    - 90mL of water
    - 10 mL of 50X TAE
        - *this provides the ions that carry the electric current through the gel. No ions, no current*
    - 0.7g of agarose

6) swirl vigorously to mix

7) microwave on high for 1.5 minutes - watch closely and stop at the first signs of boiling

8) WEAR HOT GLOVES and carefully remove the flask from the microwave. Leave the flask on the counter to cool

9) when the flask containing the agarose is cool enough to touch for 10 full seconds, add 2uL of ethidium bromide and swirl to mix
   - ethidium bromide is a carcinogen; **_you must wear gloves for this and all subsequent steps_**

10) pour the gel and be sure to include a comb as demonstrated by your instructor

---

11) view your original gel, containing your loaded samples, on the UV light box

12) while wearing gloves and eye protection, carefully use a disposable scalpel to cut your plasmid and PCR gel bands out of the gel
   - try to keep the gel band as small as possible
   - place these bands into two separate and labeled microfuge tubes
if your gel failed and you did not see bands, please obtain 'working samples' from your instructor now

GeneClean
13) add 600uL of GX1 buffer to each gel band

14) incubate the samples at 50°C, mixing frequently by inverting, until the gel is completely dissolved (this should take no more than 10 minutes)

15) be sure your column is nested into a flow-through tube. Transfer the entire sample into the spin column, and centrifuge at 1,700 RCF for 1 minute. Discard the flow through liquid.

16) add 500µl GE Buffer to spin column, and centrifuge at 15,300 RCF for 1 minute. Discard the flow through liquid.

17) add 500µl GE Buffer to spin column, and centrifuge at 15,300 RCF for 1 minute. Discard the flow through liquid. (Yes, this is a repeat of the previous step to achieve two washes of the column)

18) Centrifuge again for 2 minutes (with the column empty) to remove the residual ethanol.

19) Discard the flow-through tube. Set the column into a new, clean, labeled microcentrifuge tube.

20) to elute DNA:
- add 25 µl of TE to the column containing the PCR product DNA
- add 100uL of TE to the column containing the plasmid DNA

21) incubate at RT for 2-5 minutes

22) centrifuge samples for 30 s at 15,300 RCF

23) discard the column and be sure to label the tubes containing the purified DNA.

Ligation
24) add the following to a new clean microfuge tube:
7.5uL of purified PCR DNA of *TUB1*
0.5uL of purified plasmid DNA
1 uL of 10X ligation buffer
1 uL of DNA ligase

25) vortex both samples briefly

26) spin down for ~10 seconds at 15,300 RCF
simply to collect the entire reaction at the bottom of the tube

27) incubate at 16°C overnight
your instructor will collect and freeze your samples tomorrow

Reflection Questions:
1) Give the two different reasons for why DNA ligase requires ATP during ligation reactions.

2) Why is it beneficial for ligation reactions to be incubated at temperatures below room temperature?

Lab Exercise #5:

E. coli Transformation with Cloned TUB1-GFP Construct

Lab Exercise #5: *E. coli* Transformation

Background: If all went well in the lab exercises so far, then we've made what we set out to make: an artificial gene that encodes both the Tub1 protein from yeast (i.e., the protein subunit of microtubules) with an additional protein on the end of it: GFP. A very rough and simple schematic of the plasmid we attempted to build is shown below. By adding the DNA sequence encoding GFP to the *TUB1* gene, we will force the cell to express Tub1-GFP protein. Then, as the last lab exercise of the semester, we can see this protein in the living yeast cell.

However, today we must use *E. coli* as a DNA factory. In order to have enough *TUB1-GFP* genes to use in yeast, we must make millions of copies of it. For this, we do not want to use PCR because PCR is prone to errors and is limited in the size of the DNA it can amplify. For purposes where quality is more important than quantity and time, we use *E. coli* to copy our DNA of interest. There are proofreading capabilities of DNA polymerase in *E. coli*. For this reason, and others, we use *E. coli* as our DNA copying mechanism. The plasmid we built in our last lab exercise is specifically designed to function in *E. coli* as well as in yeast (it is referred to as a "shuttle vector" for this reason).

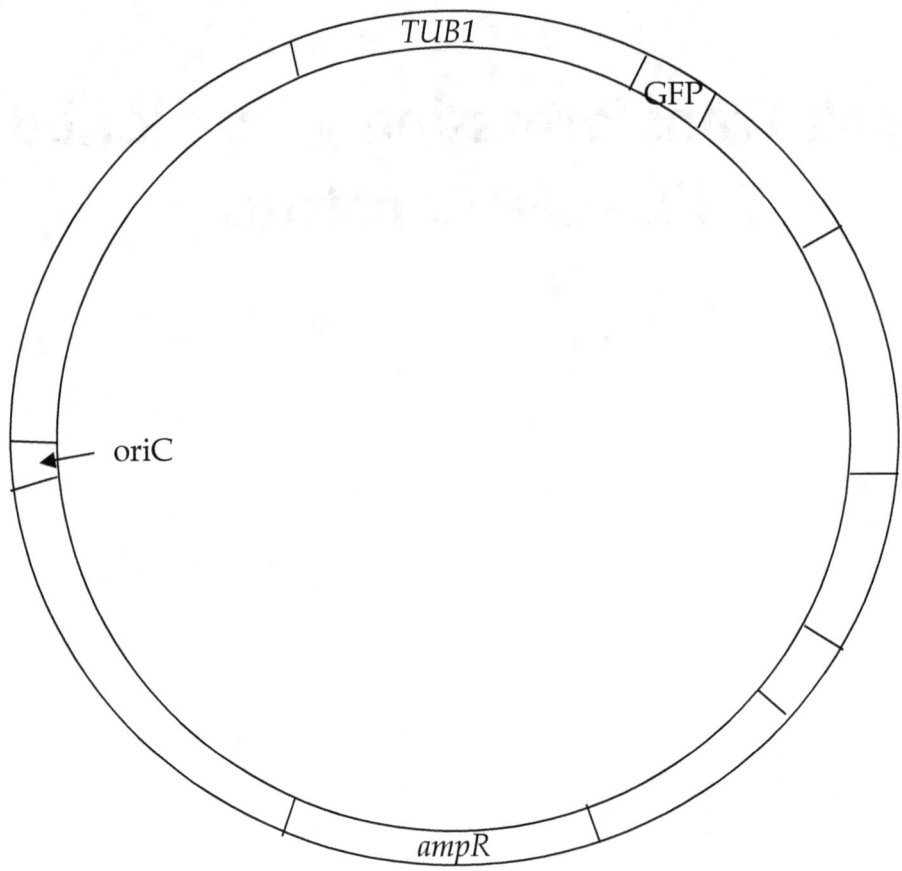

Objective: To transform *E. coli* with the plasmid encoding the Tub1-GFP protein.

Protocol

1) in the cooling incubator are tubes containing thawed *E. coli* cells that are "competent" (i.e., prepared in such a way that they are ready to receive DNA)

  - each team takes one tube of cells
  - each team takes two empty sterile microfuge tubes

2) add 50uL of cells to each microfuge tube

3) label tubes #1 and #2

4) to tube #1 add 6uL of your ligation reaction from the last lab exercise

  - to tube #2 add 1uL of the positive control DNA (this is an intact GFP plasmid *lacking* TUB1)

5) mix by gently flicking the tubes

6) place tubes in the 0°C incubator for 10 minutes

7) "heat-shock" your samples by placing at 42°C for 45 seconds (be precise in timing this step)
  the heat will make the cells porous and allow the plasmid DNA to enter the cell

8) place tubes in the 0°C incubator for 2 minutes

9) add 900uL of SOC media to each sample
  this is E. coli's *favorite 'food'. This step is essentially a rescue step. After seriously stressing the bacterial cells with high heat, you are now letting them recover. Both by providing a cold environment and also feeding them rich media.*

10) incubate at 37°C for one hour
  one hour is at least three generations for E. coli. *This step is also part of the recovery process*

11) plate 100 uL of control from tube #2 on LBA plates (Luria Broth containing ampicillin)

12) spin down tube #1 in microfuge (5 minutes at low speed)

13) pour off supernatant and resuspend cell pellet in residual media

14) plate 100uL of concentrated cells on LBA plates as in step 11

15) all plates will be incubated overnight at 37°C

Reflection Questions:
1) Without a long incubation time in Step #10, this transformation would likely fail. Why?

2) How does the amp^R gene confer ampicillin resistance? And, related, what are satellite colonies?

Lab Exercise #6:

Miniprep of TUB1-GFP Plasmid and Sequencing Reaction on the ABI3500

Lab Exercise #6: Preparation of *E. Coli* Genomic DNA ("Mini-preps") and Sequencing of the *TUB1-GFP* Recombinant Gene

Background: We used bacteria as a DNA factory; to make many many copies of our plasmid DNA. This plasmid should encode the Tub1 protein attached to a GFP tag. This new tagged Tub1 protein will be easy to visualize in later experiments. However, before we can 'see' this Tub1, we must express it (make it) in yeast. To do that we must now get our copied plasmid DNA out of *E. coli*. In principle, a genomic preparation of bacterial DNA ("mini-prep") is very similar to a genomic prep in yeast. However, in practice, mini-preps are much easier to do.

After isolating our plasmid DNA, we must be sure we have made the correct thing (i.e., TUB1-GFP). We can do this by sequencing the entire gene. To do this, we will use a fluorescence-based Sanger sequencing technique.

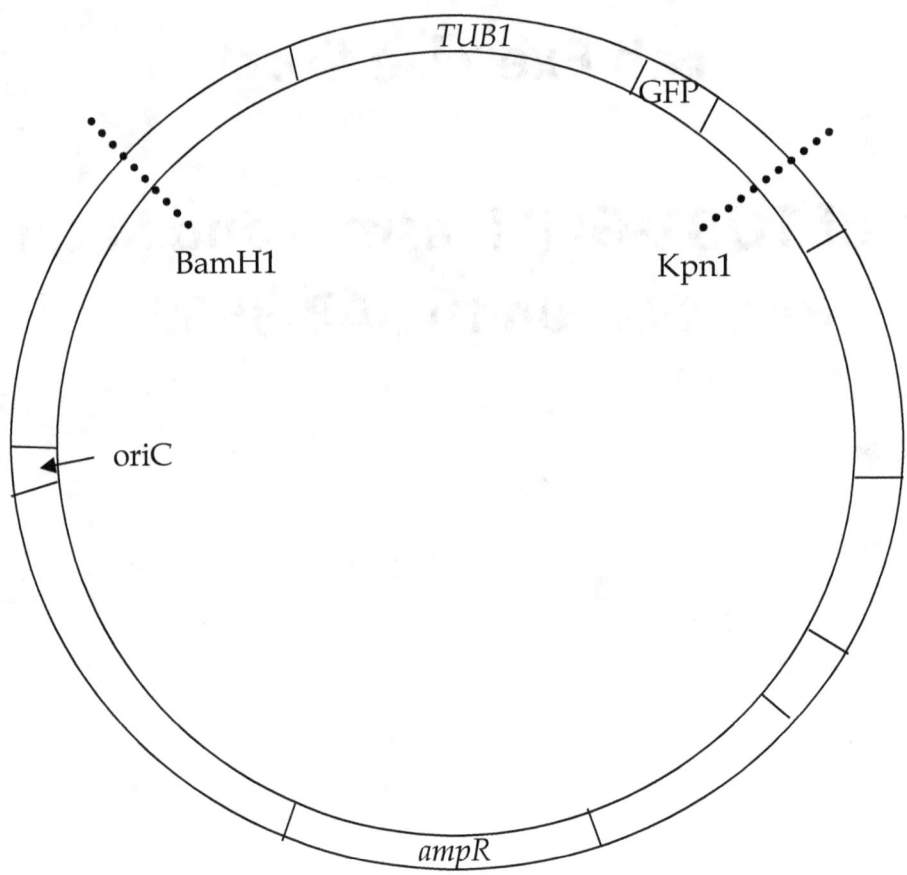

<u>Objective</u>: To prepare plasmid DNA from *E. coli* and analyze it by DNA sequencing.

<u>Protocol</u>

Yesterday E. coli *colonies were picked from your LBA plates and used to inoculate 2mLs of liquid LBA media. Cultures were grown overnight in a shaking incubator set at 37°C*

<u>*E. coli* Plasmid Mini-Prep</u>

1) transfer the cells to a microfuge tube by pouring from the culture tube into the microfuge tube. Be careful not to overflow the microfuge tube.

2) spin cells down for 5 minutes at 10,600 RCF

3) remove supernatant with a pipetteman and discard down the drain

4) add 100uL of SOLUTION I to your cell pellet and resuspend by pipetting up and down

5) incubate at room temperature on your desk for 5 minutes
   > *this step digests the cell wall of the* E. coli *cell*

6) add 200uL of SOLUTION II to your sample and mix by inverting
   > *this step dissolves the cell membrane of the* E. coli *cell with the detergent SDS*

7) incubate in the 0°C incubator for 5 minutes

8) add 150uL of SOLUTION III and mix by VORTEXING
   > *- the detergent in the lysis buffer is necessary to burst the* E. coli *cells, but it's bad news for anything else. Much like in the yeast genomic prep, potassium acetate binds to the negative detergent ions and forms a salt. This salt is insoluble and falls out of solution as a white precipitate. You'll see a white precipitate form.*

9) incubate in the 0°C incubator for 5 minutes

10) spin your samples down at 10,600 RCF for 5 minutes

11) transfer the supernatant to a clean tube (throw the white pellet away)
    - the DNA we want is still in solution – in the liquid – so KEEP THIS LIQUID!

Plasmid DNA Clean-Up

12) transfer 100uL of your sample to a new tube and add 500µl GX1 Buffer to your sample; **mix well**.

13) Obtain a column and be sure your column is nested into a flow-through tube. Transfer the entire sample into the spin column, and centrifuge at 1,700 RCF for 1 minute. Discard the flow through liquid.

14) Add 500µl GE Buffer to spin column, and centrifuge at 15,300 RCF for 1 minute. Discard the flow through liquid.

15) Add 500µl GE Buffer to spin column, and centrifuge at 15,300 RCF for 1 minute. Discard the flow through liquid. (Yes, this is a repeat of the previous step to achieve two washes of the column)

16) Centrifuge again for 2 minutes (with the column empty) to remove the residual ethanol.

17) Discard the flow-through tube. Set the column into a new, clean, labeled microcentrifuge tube.

18) To elute DNA, add 20uL of TE onto the membrane and let stand for 2-5 minutes at room temperature.

19) Centrifuge at 15,300 RCF for 1 min to elute the DNA. Discard the column.

DNA Sequencing

20) To a new, clean PCR tube add the following:
- 11uL of water
- 4uL of sequencing premix
- 2uL of sequencing buffer
- 1uL of primer
- 2uL of miniprepped *TUB1-GFP* plasmid DNA

21) Mix well and 'flick' down to bring your sample back down to the bottom of the tube

Your instructor will now place your samples in the thermocycler where the sequencing reaction will take place. The cycles will be:

An initial denaturation at 96 °C for 1 min
Repeat the following for 25 cycles:
• 96 °C for 10 sec
• 50 °C for 5 sec
• 60 °C for 4 min

The DNA will then be purified for you and run through capillary electrophoresis on the AB3500 Genetic Analyzer. The sequencing data will be collected by the instrument automatically and we will review that data together, as a class, during our next lab meeting.

Reflection Questions:
1) Solution I of the mini-prep procedure contains lysozyme. What is the role of this enzyme?

2) What is a ddNTP and what role does it play in a sequencing reaction?

Lab Exercise #7:

Analysis of Sequencing Data -
Yeast Transformation with Plasmid

Lab Exercise #7: Yeast Transformation with *TUB1*-GFP Plasmid

Background: Tubulin is a yeast protein and we would like to visualize this protein *in yeast*. Obviously, this requires that tubulin be made in yeast. So, today we will take the plasmid that we built (or the plasmid that was built for you, since some of yours might not have worked) and put it into yeast. This will give us a yeast cell that makes the GFP tagged tubulin protein that we want it to make.

The plasmid we built is called a "shuttle vector"; it can function in both bacteria and yeast. We will select for yeast cells that have successfully received the plasmid by growing our cells on media that lacks the nitrogenous base uracil. Once we get yeast cells that contain our plasmid, we then must verify that the tubulin-GFP is being made. We will do this verification over the next three weeks after this lab exercise.

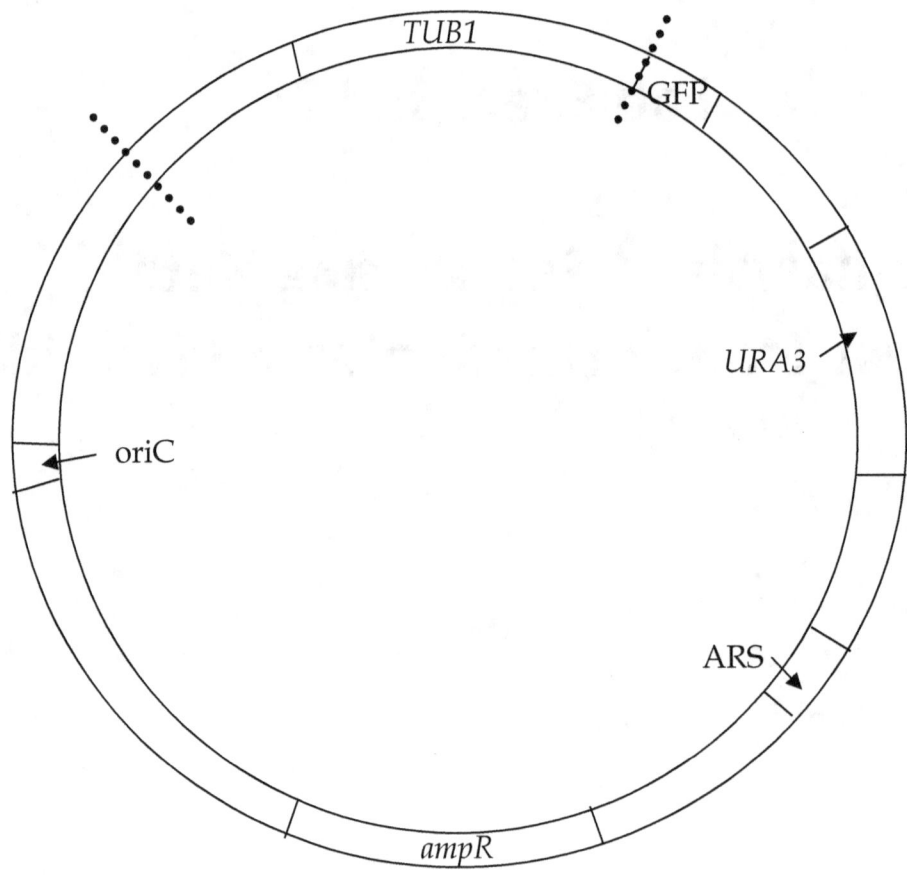

<u>**Objective**</u>: **To transform yeast cells with plasmids encoding Tub1-GFP.**

<u>**Protocol**</u>
Last night, your instructor started an overnight culture of normal, wildtype yeast cells. Before class, those cells were divided into 500uL aliquots. The rest is up to you...

1) take one microfuge tube of yeast cells

2) spin cells down for 5 minutes at 10,600 RCF

3) remove the supernatant (media) with a pipetteman
    - try to remove as much as possible

4) ***WHILE WEARING GLOVES*** resuspend the cell pellet in 100uL of "ONE-STEP BUFFER" (OSB)
    The OSB creates small holes in the cell wall and membrane of the yeast cell. These holes allow the plasmid DNA to soak into the cell. Once in the cell, good intact DNA will be shipped to the nucleus.

5) add 5uL of non-specific ssDNA to the tube of cells
    Transformations work better when there is a lot of DNA entering the cell; like an unavoidable wave hitting the cell's interior. However, our plasmid is precious and it would be a waste to use so much when only a little is needed to work in the cell. Therefore, we use non-specific, cheap DNA to achieve the 'wave' effect, and only a small percentage of that total DNA is our plasmid.

6) add 1uL of your TUB1-GFP plasmid to the tube of cells
    - ***mix by flicking the tubes***

7) incubate at 45°C for 30 minutes
    The higher temperature assists in making the cell riddled with holes. The time allows the DNA to make its way inside the cell

8) appropriately label a plate lacking uracil (-ura) for the transformation that you are doing

9) pipette cells up and down to mix and add 100uL of the transformation reaction mix to the center of the plate
    - spread the cells using glass beads

10) bring the plate up to your instructor; it will be incubated at 30°C until colonies of yeast cells become visible

Reflection Questions:

1) The one step buffer (OSB) used here contains PEG 3350. What is PEG 3350 and why is it used?

2) In Exercise #5, we required a one hour incubation for 'phenotypic lag' (Step #10 in that exercise), yet no such step is needed here. Why?

Lab Exercise #8:

TCA Whole Protein Prep from Yeast

Lab Exercise #8: Yeast Protein Preparation

Background: We're now entering the world of proteins. As we said *WAAAAAAAAAAY* back at the beginning of the course, molecular biology is primarily the world of nucleic acids (DNA and RNA), while biochemistry is primarily the world of proteins. But DNA codes for proteins, RNA is translated into proteins, and proteins make DNA and RNA – so there is a lot of overlap.

It's time to move from the *TUB1-GFP* gene onto the next (and last) steps of this semester-long project: 'seeing' Tub1-GFP protein. Our last step before visualizing Tub1-GFP *in vivo* using fluorescent microscopy is to make sure that yeast is expressing this protein. To do this, we must first get Tub1-GFP out of living cells; and, along with it, we will also extract all of the other proteins from yeast as well. That is all we will accomplish today. Just as genomic DNA preps isolate genomic DNA, protein preps are designed to isolate proteins

Objective: To prepare and collect the entire set of proteins from cells expressing Tub1-GFP.

Protocol
Previously, your instructor started overnight cultures of the Tub1-GFP cells that you all worked with in the last lab. Empty plasmid yeast cells were also prepared; these cells contain only the initial plasmid that you used, with a GFP gene, but no TUB1. All cells were spun down, washed with 20% trichloroacetic acid (TCA), and frozen at -80°C. TCA partially unfolds proteins and makes them very sticky; so much so that they stick to one another and become insoluble, falling out of solution as a precipitate.

NOTE: TCA is a weak *acid* – *ALWAYS* wear gloves today

NOTE: All steps below should be done in parallel for Tub1-GFP and empty plasmid samples

1) resuspend cell pellet in 250µl 20% TCA

2) add ~250µl glass beads and incubate at 0°C for 5 minutes
      - again, the cell wall of yeast is very durable. The vortexing with glass beads destroys this cell wall by pounding it to pieces.

3) vortex three times for 1 minute each time

4) carefully remove as much of the sample as possible with a pipetteman and transfer to a new, clean tube (*your instructor will demonstrate*)

5) add 300µl 5% TCA to beads to wash them

6) remove this 5% TCA from beads and pool with sample from step #4

7) add 700µl of 5% TCA to your sample (giving a ≈1.25ml final volume) and pipette up and down to mix

8) spin at 20,800 RCF for 10 minutes

9) discard the supernatant and wash pellet with 750µl of ice-cold 100% EtOH
   - pipette up and down and spin for 5 minutes at 20,800 RCF
   - *this dehydrates the protein pellet removing all traces of TCA*

10) discard all of the supernatant and resuspend pellet in 40µl 1M Tris (pH8.0)
    - pipette up and down to resuspend

11) add 80µl 2xSDS Loading Buffer
    - *this loading buffer contains DTT (smells like rotten eggs) be sure your gloves are on and be prepared for the stink. Use the hood if you'd prefer.*

12) incubate for 8 minutes at 95°C – **BE CAREFUL; THE HEAT BLOCK IS VERY HOT!**
    - *the SDS and DTT in the loading buffer begins denaturing the proteins (unfolds them). The heat finishes the job leaving proteins in their primary and secondary structures. These proteins do not refold due to the SDS molecules coating the proteins. SDS also confers a negative charge making these proteins behave like DNA in gel electrophoresis.*

13) spin at 20,800 RCF for 5 minutes

14) transfer supernatant to new microfuge tube and discard the debris pellet

15) bring samples up to your instructor so that they can frozen at -20°C

Reflection Questions:
1) How do Steps #11 and #12 reverse the effect of TCA and re-solubilize the yeast proteins?

2) Why are glass beads used to lyse the cells here, instead of zymolyase (as we used in Exercise #1)?

Lab Exercise #9:

SDS-Polyacrylamide Gel Electrophoresis (PAGE) of Whole Protein Extract

Lab Exercise #9: Western Blot, part I (SDS-PAGE and transfer)

Background: This week we will begin the process of validating the expression of our Tub1-GFP protein. Today, we will size fractionate our protein samples. Boiling in the presence of negatively charged SDS (part of the loading buffer added last week) unfolds the protein and coats it in negative charge. In very crude terms, your proteins now resemble DNA: long linear molecules of negative charge. These denatured proteins will travel through a polyacrylamide gel much as DNA traveled through an agarose gel. This technique is referred to as SDS-PAGE (for "**p**oly**a**crylamide **ge**l **e**lectrophoresis" in the presence of SDS). Longer proteins will move slower; smaller proteins faster. Once the gel is run, the proteins will be transferred to a nitrocellulose or PVDF membrane. This is a sheet of material that proteins stick to readily. This transfer creates an imprint of the gel (with proteins separated by size), but these proteins are now accessible (in the gel, the proteins could not be accessed, bound, or recognized). Once the proteins are on the membrane, they can be recognized and bound by antibodies. That is where we will pick up this experiment the next time we meet.

Objective: To separate proteins based on size and transfer to membrane as part of a western blot.

Protocol
Unpolymerized acrylamide (in liquid form) is a neurotoxin. As such, pre-cast SDS-PAGE gels are typically purchased from a commercial vendor. However, you will do the rest: load your samples, run the gel, and set up the transfer.

1) spin down your samples – 4 minutes at 10,600 RCF

2) load Tub1-GFP and empty plasmid samples on SDS-PAGE gel
   – 30uL of each sample per well

MAKE NOTE OF WHERE YOU LOADED YOUR SAMPLES AND THE ORDER THEY ARE LOADED IN

3) run at 150V for ~1 hour

Transfer (two groups will team up together per transfer/gel)
4) soak two sponges, two sheets of filter paper and one sheet of nitrocellulose membrane in cold transfer buffer

5) take gel apart as shown by your instructor

6) assemble transfer on the negative/black side of cassette

7) sponge – filter paper with gel facing UP – nitrocellulose membrane (the white 'paper', not the blue!) – filter paper – sponge (make the 'sandwich')

8) close cassette and place in chamber in the correct orientation so that the negative electrode is closer to the gel and the positive electrode is closer to the membrane – your instructor will explain/demonstrate

9) add frozen pack and stir bar

10) fill chamber with cold transfer buffer
     fill completely _without_ spilling

11) make sure red wires connect to red electrodes and black wires to black electrodes

12) begin transfer at 100V (it will continue for 1.5 hours)

13) your instructor will do the rest (disassemble the transfer and store the blots in TBS)

Reflection Questions:
1) There is glycine and SDS in the running buffer used for running this PAGE gel. What is the role of each of these reagents?

2) Western blotting is also often referred to as electroblotting. Why might this be?

Lab Exercise #10:

Western Blot for Tub1-GFP (using GFP antibodies)

Lab Exercise #10: Western Blot, part II (developing)

Background: Today we will finish our western blot and (hopefully) visualize our protein band for Tub1-GFP.

Yesterday, your instructor 'blocked' your blots with non-specific proteins to saturate the membrane with protein. Then, two hours later, your instructor added the primary antibody, which directly recognizes and binds to the GFP region of your protein. This was left to incubate at room temperature until now. This is where you will pick up the protocol and finish the experiment.

Objective: To determine the presence of Tub1-GFP by western blotting.

* WEAR GLOVES AT ALL TIMES DURING THIS LAB EXERCISE *

Protocol

1) transfer the solution covering the blots to an empty 50mL conical tube
    - this is the solution containing blocking protein and the primary antibody to GFP

2) rinse the blot with tris buffered saline (TBS), dump TBS down the drain

3) wash the blots – THREE TIMES – by adding TBS and incubating on the 'shaker' for 3 minutes each wash (TBS always goes down the drain)

4) add the secondary antibody
    - *this is the antibody which will recognize and directly bind to the primary antibody. This is also the antibody which is covalently bound to an enzyme (alkaline phosphatase) and will catalyze a color change for us when we add substrate in the last step*

5) incubate on the 'rocker' for ~one hour

6) dump the secondary antibody down the drain; rinse the blot with TBS, dump TBS down the drain

7) wash the blots – THREE TIMES – by adding TBS and incubating on the 'shaker' for 3 minutes each wash (TBS always goes down the drain)

8) add Western Blue substrate and wait to see the color change at a single protein band
    - this step may take up to 15 minutes...
    - be sure to snap a picture; *these are results!*

Reflection Question:
1) What would be the expected consequences if your instructor forgot to saturate your blot with milk proteins before adding the primary antibody?

Lab Exercise #11:

Fluorescent Microscopy to Visualize Tub1-GFP

Lab Exercise #11: Fluorescent Microscopy to Visualize Tub1-GFP

Background: We've made it! Today, we will see our yeast cells fluoresce green as a result of the microtubules of their cytoskeletons being labeled with green fluorescent protein. If we can achieve a high enough magnification and resolution, we may even see some of the structure of the microtubules/cytoskeletons itself.

Congratulations on making it through this semester-long research project. Well done!

Objective: To visualize Tub1-GFP recombinant protein in yeast using fluorescent microscopy

Protocol

Your instructor will demonstrate the preparation of your yeast slides and the use of the fluorescent microscopy platform. Pay close attention and take notes as needed.

Instructors' Notes

Lab Exercise #1: Yeast Genomic DNA Preparation – Instructor's Notes

One day prior to the lab exercise, start overnight cultures of freshly streaked wildtype yeast cells in rich liquid media (e.g., YPD), shaking at 30°C. Grow approximately 2mL of cells per group/student.

On the morning of the lab exercise, spin these cells down at low speed (~3800 RCF) for five minutes. Remove the supernatant and wash cell pellet with ~25mL of zymolyase buffer and repeat the spin down. Resuspend cell pellet in one-fourth the original volume of the overnight culture in zymolyase buffer. Aliquot as necessary.

Solutions:

YPD (per liter):
10g yeast extract
20g peptone
20g glucose
(15g agar; for plates only)
autoclave

zymolyase buffer:
1M sorbitol
100mM EDTA
pH 7.5

zymolyase:
10mg/mL in water

lysis buffer:
50mM Tris-HCl pH 7.5
20mM EDTA
1% SDS

5M Potassium Acetate

The solutions from steps 13 through 20 refer to those provided as part of the GelElute Extraction Kit from 5Prime (cat. no.: 2300400)

Lab Exercise #2: PCR and Gel Electrophoresis of *TUB1* – Instructor's Notes

To expedite progress through this exercise, especially due to the 'down time' of the PCR itself, you may want to prepare completed PCR amplifications for each student/group ahead of time. These samples can be substituted for the students' PCRs once the students have completed the set-up of their reactions, allowing students to run their gels on the same day as the PCR set-up. Please note: students can always elect to run their own PCRs on a gel at a later date to see if their reactions were successful.

Solutions:

TUB1 Primer Sequences:
Forward: GAC TGG ATC CGA GTC TAT CAA TGG CGG GCA CTG CCA TCC G
   - anneals at -750 and contains a BamH1 site
Reverse: GAC TGG TAC CAA ATT CCT CTT CCT CAG CGT ATG AGT CGG C
   - anneals at +1457 and contains a Kpn1 site

All primers should be stored at 100uM storage stocks, and then diluted to 10uM working stocks and used as per Taq polymerase manufacturer's recommendations.

For all other PCR reagents, use the manufacturer/vendor of your choice and follow their recommendations for reaction and cycling conditions.

10X TAE (per liter):
48.5 g Tris
11.4 mL glacial acetic acid
20 mL 0.5M EDTA
pH 8.0

All gels are 0.7% agarose in 1X TAE. DNA may be stained as per the instructor's preference.
   - please note, step #10 is specific for ethidium bromide DNA staining

5X agarose gel sample buffer:
5mL 10X TAE
2mL water
3mL glycerol
2.5mg bromophenol blue
2.5mg xylene cyanol

Lab Exercise #3: Restriction Digests to Prepare for Cloning – Instructor's Notes

The agarose gel electrophoresis from the previous step is used diagnostically only. Students purify the *TUB1* amplicon from the remaining ~30uL of PCR sample directly. Choosing to purify the entire PCR sample via gel electrophoresis would require some modification of the protocol below.

Solutions:

Steps 2 through 9 refer to the same reagents as in Exercise #1 (i.e., GelElute Extraction from 5Prime), but with the use of generic DNA capture columns.

TE:
10mM Tris-HCl pH 8
1mM EDTA

plasmid DNA:
The GFP plasmid used by the author is no longer commercially available. There are a number of suitable GFP cloning vectors currently available that would suffice for this exercise. However, be sure to check the BamH1 and Kpn1 RE sites used here to ensure that they are (1) compatible with the MCS of the vector you choose and (2) in-frame with the GFP gene in that vector. Also, be sure to use a shuttle vector that not only contains an *E. coli* oriC and selectable marker, but also a yeast ARS and auxotrophic marker.

Lab Exercise #4: Gel Purification and Ligation of *TUB1* and GFP Vector – Instructor's Notes

As part of this exercise, students practice pouring their own gels. These gels are 0.7% agarose in 1X TAE, as above, with DNA staining at the preference of the instructor. I recommend having a gel prepared ahead of time, however. Students often struggle with pouring a useable gel and having students load their samples on to the prepared gel prior to pouring their own also expedites the exercise. When multiple lab sections are meeting, back-to-back, on the same day, successful gels from earlier sections can be used as the prepared gel for later section. This eases the instructor's burden, and gives students of earlier sections a greater sense of accomplishment.

Both cloning fragments, from the *TUB1* and GFP vector digests, are purified from the agarose gel to isolate them from the 'scrap' fragments, which include compatible sticky ends. Here again, students use reagents from the GelElute Extraction Kit from 5Prime coupled with generic DNA capture columns.

I recommend having working, digested and gel purified samples of both *TUB1* and vector prepared ahead of time (gel bands can be safely frozen at -20°C for 1-4 days). These working samples can be provided to students/groups that do not see any bands on their gels after step 12.

Solutions:

All gel electrophoresis reagents are as for previous exercises

All DNA purification reagents refer to those provided with the GelElute Extraction Kit from 5Prime

Ligase: any generic T_4 DNA Ligase should do

Lab Exercise #5: *E. coli* Transformation with Cloned *TUB1*-GFP Construct – Instructor's Notes

We use competent HB101 *E. coli* for these transformations (available through Promega; cat. no.: L2015). These cells are divided into ~100uL aliquots prior to the lab exercise beginning.

The positive control plasmid referred to in step #4 is an intact GFP plasmid; the same as was used in the digestion for Exercise #3.

Solutions:

SOC Media (per liter):
20g Tryptone
5g Yeast Extract
2ml of 5M NaCl
2.5ml of 1M KCl
10ml of 1M MgCl2
10ml of 1M MgSO4
20ml of 1M glucose
autoclave

LBA plates (per 950mL):
5g Yeast Extract
10g Typtone
10g NaCl
1g Glucose
adjust pH to 7.5 with NaOH
15g Agar
autoclave
allow to cool to ~65°C
add 50mL of filter sterilized 1mg/mL ampicillin

Lab Exercise #6: Miniprep of TUB1-GFP Plasmid and Sequencing Reaction on the ABI3500 – Instructor's Notes

Here, the miniprep is used to isolate the plasmid from the *E. coli* cells. This is done 'old-school' with miniprep solutions made in-house.

The instructor must start overnights of students' *E. coli* transformants the evening before this exercise. These should be 2mL overnights in LBA (the same recipe as in Exercise #5 without the agar added) grown with shaking at 37°C.

The second portion of the exercise involves sequencing the cloned insert to validate it is correct. My home institution is fortunate enough to have an ABI 3500 Genetic Analyzer as part of its forensic biology program. Therefore, we do this sequencing here ourselves. Without that capability, representative clones could be sent out for sequencing or diagnostic restriction enzyme digestions could be done to verify that predicted fragments are generated based on the correct clone. Please note: all the sequencing reaction steps and parameters described below are specific for the 3500 and the BigDye sequencing kit purchased from Life Technologies. Protocols for running sequencing reactions on the 3500 can be found via Life Technologies support documentation. Any other sequencing platforms or strategies would require extensive modifications to steps 20 and 21 below.

Solutions:
Solution I:
50mM Glucose
10mM EDTA
25mM Tris-HCl pH 8
autoclave
4mg/mL lysozyme

Solution II:
0.2M NaOH
1% SDS

Solution III (per 100mL):
60 mL 5M Potassium Acetate
11.5mL Glacial Acetic Acid
28.5mL water

Here again, students use reagents from the GelElute Extraction Kit from 5Prime coupled with generic DNA capture columns; this is the last DNA purification step of the course.

All sequencing reagents refer to those provided as part of the Life Technologies BigDye Terminator v3.1 Cycle Sequencing Kit. The sequencing primer can either be the universal primer compatible with your GFP vector or the forward *TUB1* primer used in Exercise #2.

<u>Lab Exercise #7: Analysis of Sequencing Data - Yeast Transformation with Plasmid –</u>
<u>Instructor's Notes</u>

The analysis of sequencing data component of this exercise is not included in the protocol below. Students are simply provided with the known sequences for *TUB1* and GFP and provided with the results of their sequencing reactions. They are prompted to compare the two to validate that their construct is correct. It is not uncommon for student sequencing reactions to have failed. If this occurs, students are given a choice: continue on with their construct even though its validity is uncertain, or substitute a known, working *TUB1*-GFP fusion construct that can be trusted.

One day prior to the lab exercise, start overnight cultures of freshly streaked wildtype yeast cells in rich liquid media (e.g., YPD), shaking at 30°C. Be sure these yeast cells are auxotrophic *ura3* mutants. Grow approximately 1mL of cells per group/student and divide into ~500uL aliquots.

Also, be sure to prepare synthetic complete media plates lacking uracil (or lacking the appropriate component given the auxotrophic mutant/marker you are using for your plasmid selection).

<u>Solutions:</u>
<u>One-Step Buffer (OSB):</u>
0.2M Lithium Acetate
40% PEG 3350
100mM DTT (added fresh from 1M stock just before use)

<u>ssDNA:</u>
1 mg/mL sheared and denatured salmon sperm DNA in 1XTE

<u>Complete Media lacking uracil (CM-ura) plates</u> (per liter):
5mL 200X amino acid mix
2mL 2% histidine
2mL 2% leucine
2mL 2% tryptophan
40mg adenine
40mg tyrosine
20g glucose
6.7g yeast nitrogenous base without amino acids (BD Difco; cat. no.: 291940)
15g agar
autoclave

<u>200X Amino Acid Mix</u> (per 500mL):
1g arginine
6g isoleucine
4g lysine
1g methionine
1g threonine
6g phenylalanine
autoclave

Lab Exercise #8: TCA Whole Protein Prep from Yeast – Instructor's Notes

Prior to this lab exercise, start overnight cultures of each group's/student's transformed yeast cells in CM-ura media (follow the recipe provided in Exercise #7, leaving out the agar), shaking at 30°C. Grow approximately 2mL of cells per group/student. Spin those cells down (10,600 RCF; five minutes) and resuspend cell pellets in an equal volume of cold 20% trichloroacetic acid (TCA). Repeat the spin and remove the supernatant. Freeze cell pellets at -80°C. Also grow a large volume of yeast cells harboring an empty vector, preferably the same vector backbone as the shuttle vector containing *TUB1*-GFP. Wash these cells in cold 20% TCA and divide and freeze so that each student/group will also have an 'empty vector' sample. This will serve as the negative control moving forward. Students pick up from here, using these pre-frozen cell pellets.

Solutions:

100% TCA:
100g TCA
45.4 mL water

use this as the stock solution for making the 20% TCA and 5% TCA dilutions

2X SDS-PAGE Sample Buffer:
60mM Tris-HCl pH 6.8
2% SDS
10% Glycerol
0.2% bromophenol blue
100mM DTT (added fresh from 1M stock immediately before use)

Lab Exercise #9: SDS-Polyacrylamide Gel Electrophoresis (PAGE) of Whole Protein Extract – Instructor's Notes

Students will be running their whole protein extracts on SDS-PAGE gels. These gels can be purchased pre-cast, or cast ahead of time by the instructor. I strongly recommend undergraduates not be allowed to pour acrylamide-containing gels themselves. Also, be sure to use a gel electrophoresis system that is compatible with electro-blotting. This gel will be immediately followed by western blotting on to nitrocellulose (or PVDF) membrane.

Solutions:

Stacking Gel:
6.6mL water
2.5mL 0.5M Tris-HCl pH 6.8
1mL 45% acrylamide
50uL 20% SDS
to polymerize:
100uL 10% ammonium persulfate
10uL TEMED

Resolving Gel:
5mL water
2.5mL 1.5M Tris-HCl pH 8.8
2.7mL 45% acrylamide
50uL 20% SDS
to polymerize:
100uL 10% ammonium persulfate
10uL TEMED

5X Running Buffer:
120mM Tris Base
1M Glycine
0.5% SDS
pH should come to be ~8.3

Transfer Buffer:
120mM Tris Base
1M Glycine
20% methanol
store ice-cold

TBS – pick your favorite recipe

Lab Exercise #10: Western Blot for Tub1-GFP – Instructor's Notes

One day prior to this exercise, block the students' membranes. This is typically done with 5% powdered skim milk in TBS, but some prefer to use BSA, instead. Allow the blocking step to proceed for at least 2h at room temperature. Then, remove the blocking solution and replace with primary antibody. The primary antibody should be in blocking solution and specific for GFP. Use the antibody manufacturer's/vendor's recommendations regarding proper dilution for western blotting. Allow the primary antibody to remain on the blots overnight. Students will pick up from here.

Regarding the secondary antibody, while the instructor must choose a secondary antibody that will recognize and bind to the primary, it is critical that the secondary antibody also be conjugated to alkaline phosphatase. Students will be visualizing their protein bands with the chromogen reagent, Western Blue (Promega; cat. no.: S3841) which requires AP to react.

Solutions:
TBS – choose your favorite recipe

Lab Exercise #11: Fluorescent Microscopy to Visualize Tub1-GFP – Instructor's Notes

All fluorescent microscopy platforms are different, and as such the protocol below is brief. GFP emits at the standard green fluorescence wavelength and typically requires a FITC filter. Also, many coverslips block some/all of the excitation wavelength and so much better results are typically observed with dry mounts of cell smears. However, each instructor will have to troubleshoot his/her own platform(s) to achieve the best results and then instruct and demonstrate those practices for the students.